BEFORE DISASTER STRIKES

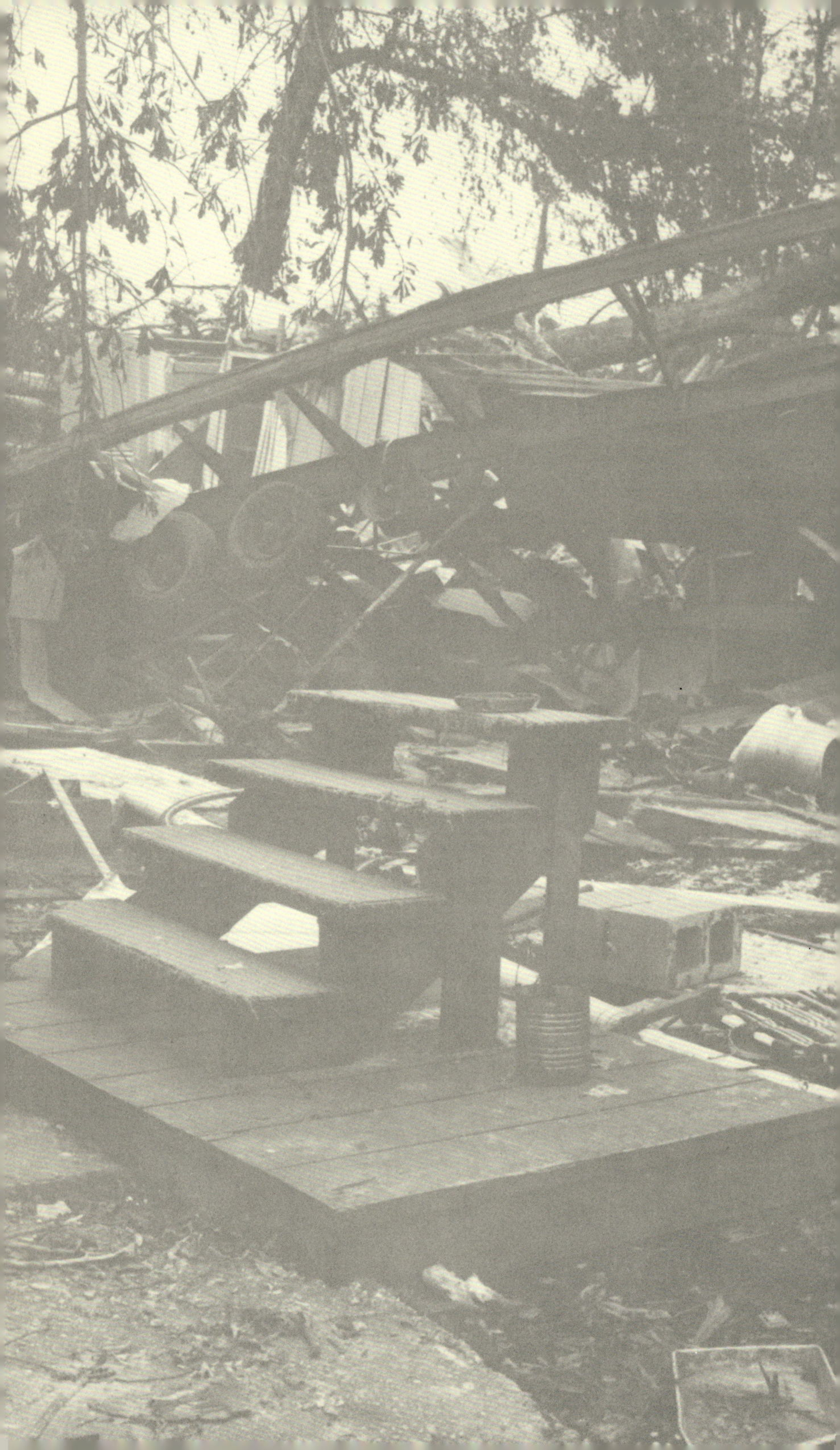

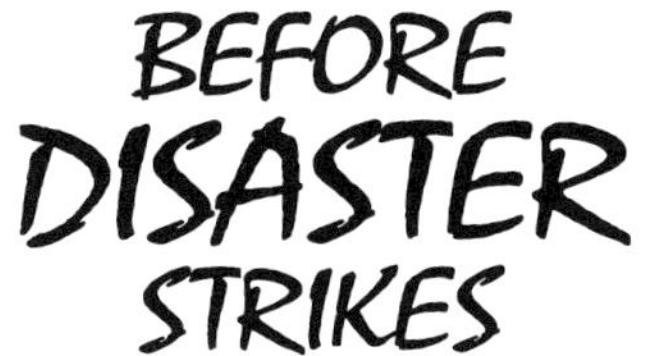

PREVENTION, PLANNING, AND RECOVERY

Caring for Your Personal Collections
In the Event of Disaster

by Priscilla O'Reilly Lawrence

Photographs by Jan White Brantley

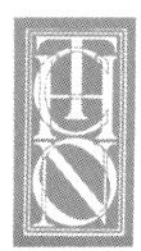

The Historic New Orleans Collection
1992

Cover, High water from Hurricane Betsy, 1965. G. E. Arnold.
THNOC acc. no. 1974.25.11.75
Half-title page, Hurricane Betsy evacuees at Capdau School,
New Orleans, Louisiana, September 1965. Courtesy the New Orleans
Times-Picayune
Frontispiece, Tornado damage at museum of A. Montz Co., La Place,
Louisiana, 1992. Jan White Brantley
Facing page 1, Cabildo fire, New Orleans, Louisiana, May 11, 1988.
Jan White Brantley
Back cover, Waterspout on Bay St. Louis, Mississippi, September 13,
1953. John Lambert III. Courtesy the New Orleans *Times-Picayune*

The Historic New Orleans Collection, New Orleans, Louisiana 70130
©1992 by the Historic New Orleans Collection
All rights reserved
Printed in the United States of America

Design: Michael Ledet Art & Design, New Orleans
Typography: Cox Electronic Publishing
Typeface: Goudy
Printing: Harvey Press

Library of Congress Cataloging-in-Publication Data

Lawrence, Priscilla O'Reilly
Before disaster strikes : prevention, planning, and recovery : caring for your personal collections in the event of disaster / by Priscilla O'Reilly Lawrence.
48 p. 22 cm.
Includes bibliographical references.
ISBN 0-917860-32-2 (pbk. : alk. paper)
1. Cultural property, Protection of – United States. 2. Personal paraphernalia – Protection – United States. 3. Art – Private collections – United States. I. Title.
N5215.L38 1992
702'.8'8–dc20 92-37511
CIP

CONTENTS

Introduction 1
Prevention 3
Planning 9
Recovery 13
Paintings 13
Paper and Books 17
Photographic Materials 22
Textiles 25
Furniture and Wooden Objects 28
Metal Objects 32
Glass and Ceramics 34
Where to Find Conservation Information 37
Bibliography 39

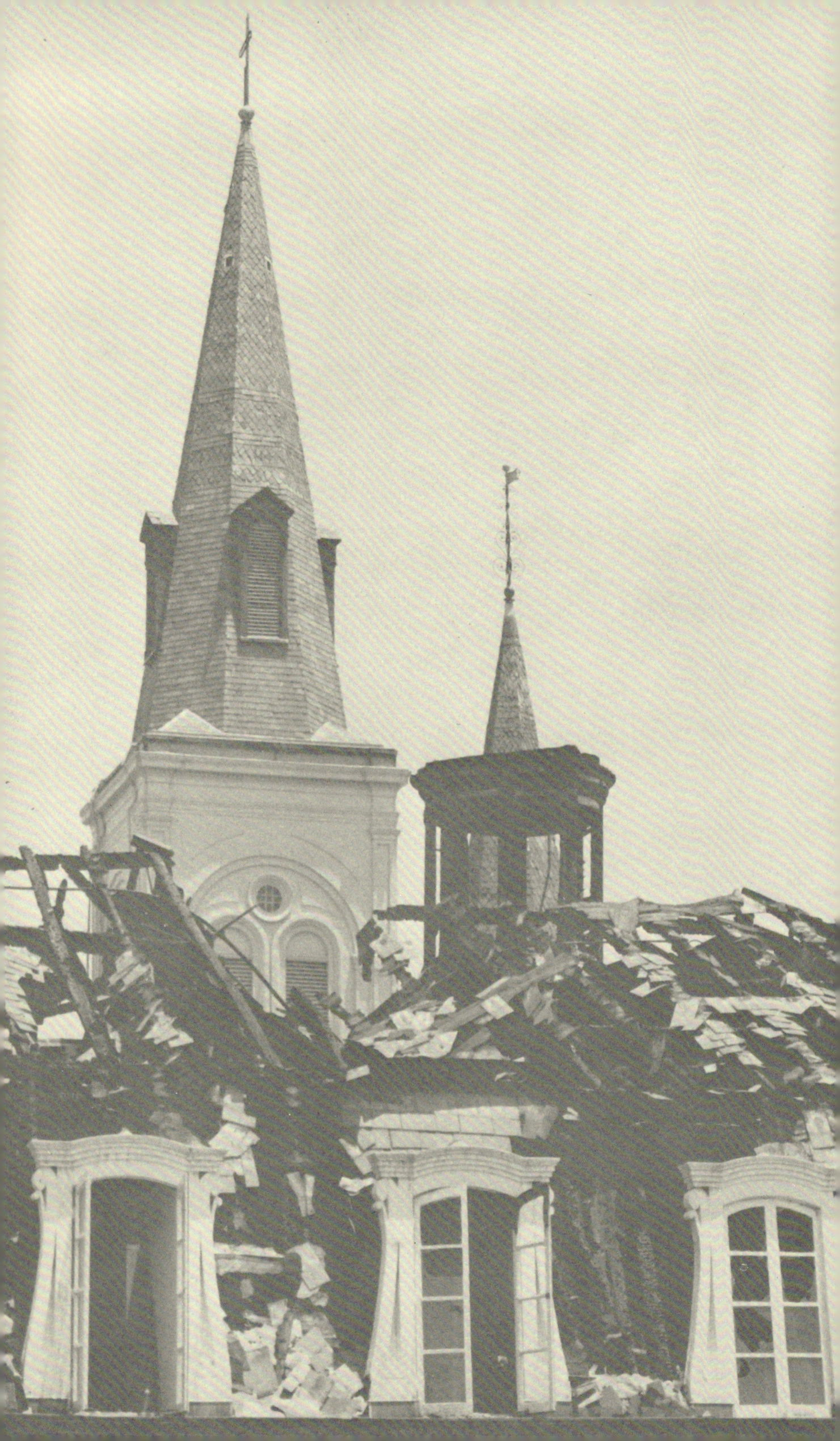

On May 11, 1988, the people of New Orleans learned that fire was spreading through the upper floor of the Cabildo, one of the Louisiana State Museum's incomparable properties. As a horrified city watched, an entire storage area filled with antique and Louisiana-made furniture went up in smoke. The building itself, erected by the Spanish colonial government in 1795, was threatened with total destruction. Thanks to the efficiency of the museum staff of LSM and the New Orleans Fire Department, other collections were saved from flames and severe water damage. The building itself, for the most part, was spared although the top floor was a total loss. On September 21 of the following year, the winds of Hurricane Hugo, recorded at 135 miles-per-hour, slammed into Charleston, South Carolina. A few weeks later, on October 17, a major earthquake rocked San Francisco.

Shocked by the Cabildo fire and two natural disasters of epic proportions in such a short time, the museum community began a serious reconsideration of its state of preparedness for dealing with such threats. Reflecting also on the flood caused by plumbing construction at the Chicago Historical Society in 1987 further prompted many museums to review their emergency procedures. In light of these recent catastrophes, museum professionals attended disaster-preparedness workshops throughout the country to learn how other institutions have responded in a crisis. Hurricane Andrew, roaring through Florida and Louisiana in August 1992, underlined the need for planning and preparedness.

Before Disaster Strikes complements the series of *Preservation Guides* published by the Historic New Orleans Collection. This volume, like the guides, offers information to the public about the dangers that can befall valuable possessions and appropriate means to safeguard family heirlooms. It also suggests steps to be taken for recovery when damage cannot be avoided.

When preparing for an emergency, the first concern must be for the family's well-being. No object, however rare, is worth the slightest risk to a person's safety. But once you have a plan for protecting yourself and your family, you can consider the best way to prevent damage to your valuables. Advance planning can prevent a great deal of damage. Even when catastrophe is inevitable, the same planning can mitigate your loss and lead to reasonable recovery.

PREVENTION

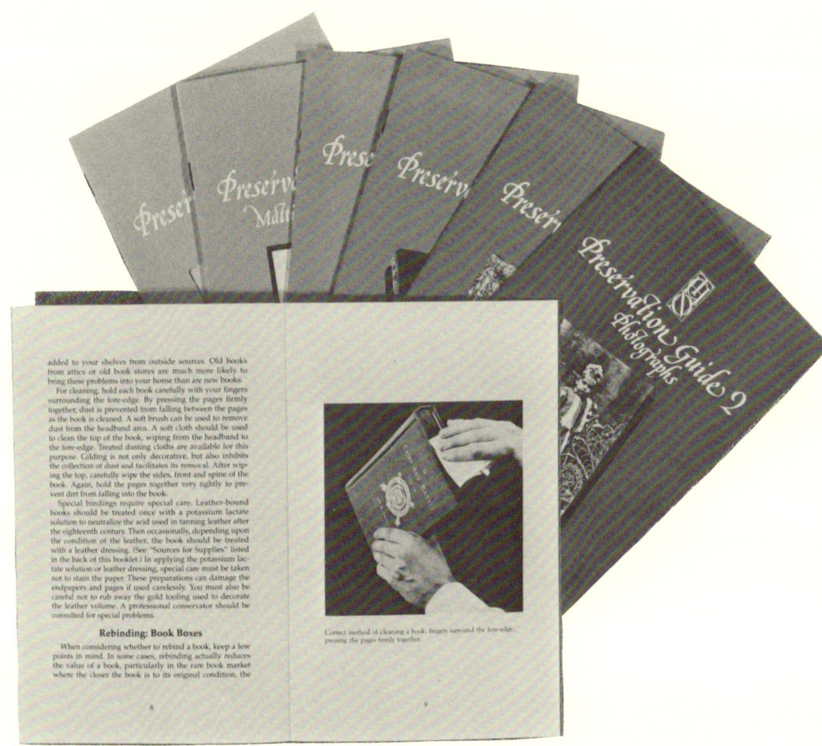

Most damage to valuable objects results from an improper environment or from careless handling. Proper basic care will prevent minor damage and most disasters. The Historic New Orleans Collection's *Preservation Guides* outline fundamental methods for the care and handling of family papers and other important documents, photographs, paintings, furniture, and books. Understanding how fluctuation in temperature and humidity causes damage to art, antiques, manuscripts, and books is the first step in prevention.

Maintaining constant temperature and humidity levels is essential. High heat and humidity, besides causing swelling, promote the growth of mold and mildew on most materials; with metals, there is the danger of rust. If an object dries out too quickly, wood will split, veneer will pop off, paint may crack and peel, and paper and textile fiber will deteriorate.

RISK ASSESSMENT

Once you have created a safe environment for your possessions, prevention of damage caused by catastrophe is the next step. It is important to determine potential risks to your possessions. First consider the kinds of natural disasters most likely to occur in your region. Your area may be periodically affected by earthquakes or tornadoes, or you may live in coastal areas that are prone to flooding or susceptible to hurricanes. Other types of natural occurrences endanger all regions. Information can be obtained from police and fire departments, state and local civil defense offices, and from the U. S. Geological Survey. Your insurance company can also provide information.

The following is a list of possible dangers:

Natural Disasters

Severe rainstorm
Electrical storm
Wind storm, tornado
Hurricane
Dust storm
Freezing precipitation
Flash flood
Slow-rising water
Volcanic eruption
Earthquake

Besides dramatic weather hazards, other common incidents can cause disaster. In assessing potential risks, it is wise to consider these as well.

Technological or Mechanical Hazards

Structural collapse	Water supply failure
Explosion	Sprinkler-system malfunction
Fire	Electrical-power failure
Air-conditioning failure	Extreme air pollution
Heating failure	Fuel or chemical spill
Humidity-control failure	Structural leaks

Human Hazards

Theft
Vandalism
Mishandling

You should make a general assessment of your vulnerability to damage or loss. A routine inspection of your home can help to identify existing hazards. **Remember that fire is the single most common and destructive disaster threatening your possessions.** Your local fire department can help you evaluate the basic safety of your home or office, recommend proper home detection devices and extinguishers, and make sure you know how to use them. In many cities, a representative of the fire department is available to come to your home. Smoke detectors and fire extinguishers are readily available and will measurably lessen the risk of serious fire damage. Sprinklers or other fire suppression systems, though costly, are the best of all options.

Other measures can help to prevent major damage from disasters. Security alarm systems can protect against theft and vandalism. Exterior window shutters can be closed during storms and will protect from heavy wind and precipitation. Valuables may be permanently stored or displayed away from window areas. Precious objects can be kept above floor level if floods are likely. Breakables can be secured on display in earthquake-prone areas with specially constructed stands and supports, and large,

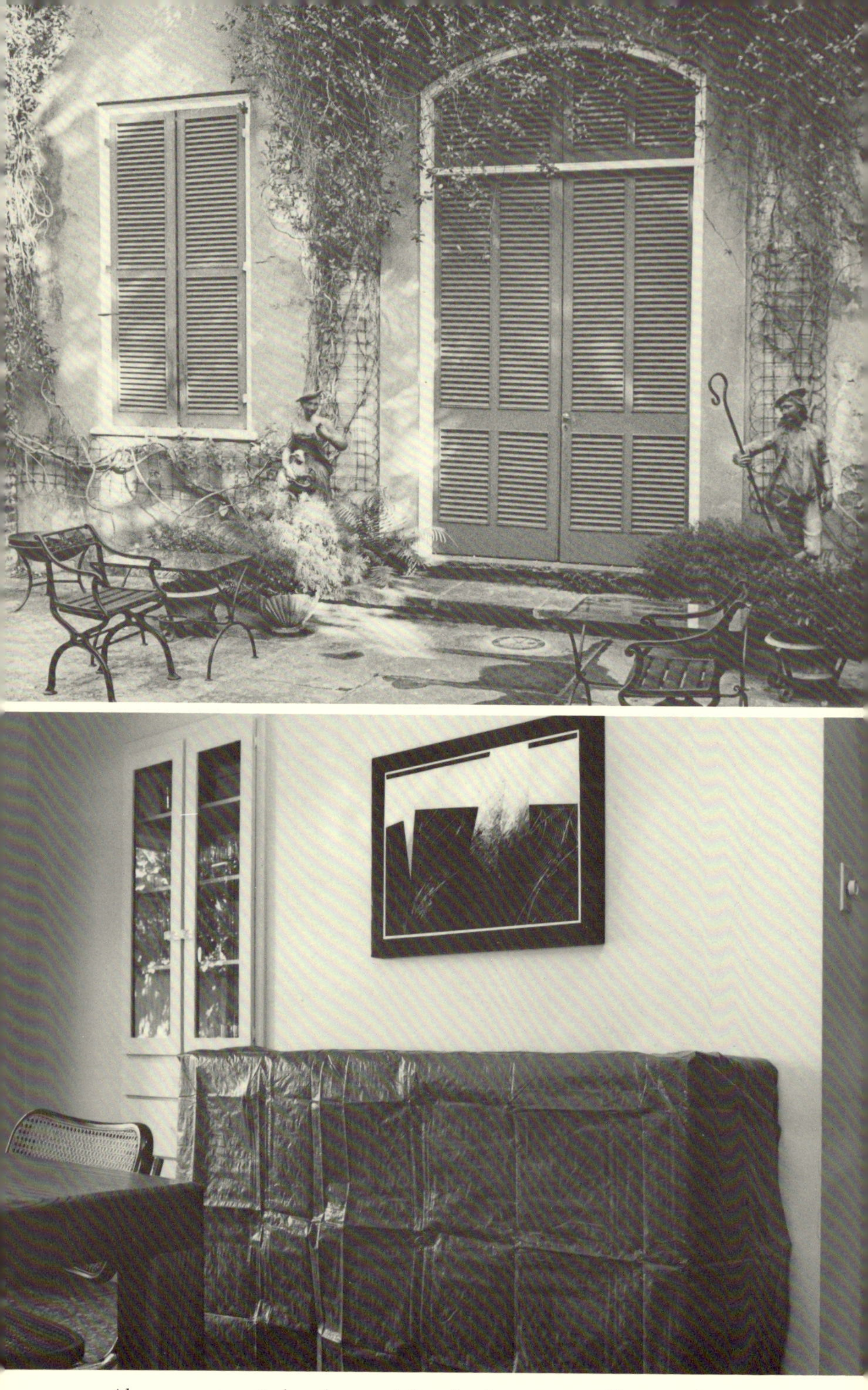

Above, exterior window shutters; *below*, furniture protected by tarpaulin

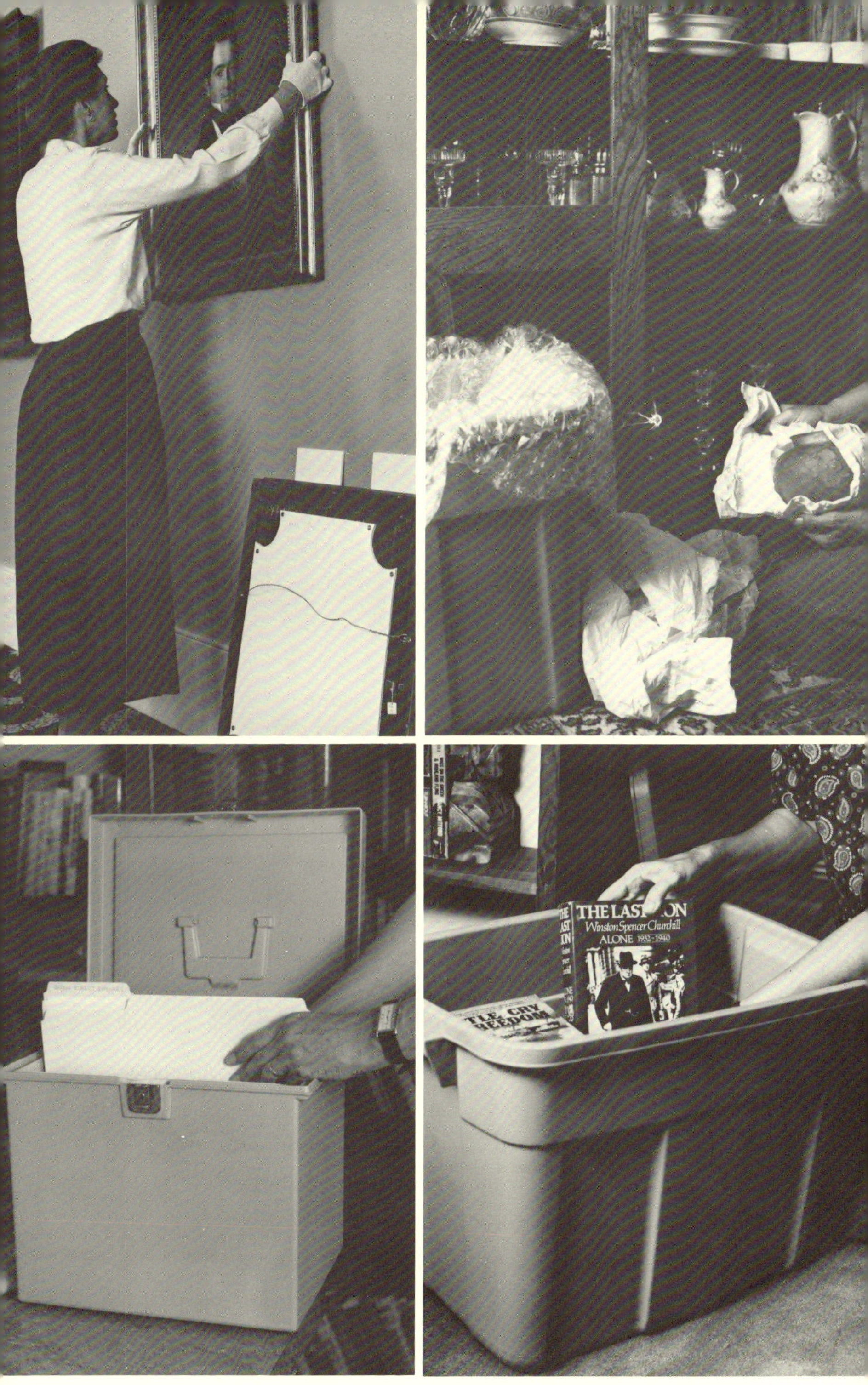

Above left, removing paintings from the wall, *right*, packing fragile items in a padded container; *below*, placing documents and books in waterproof containers

top-heavy pieces of furniture can be secured to walls. You should also carefully supervise any workmen on your premises to guard against careless handling and theft.

Relocating possessions before the occurrence of natural disasters is often impossible, but there may be time for a certain amount of preparation. With advance notice of a storm, move objects away from windows to the more structurally sound center of the building. Check the security of objects hanging on the walls and take down paintings and mirrors. They can be placed on the floor leaning face in against an interior wall, provided there is no possibility of flooding. Place breakables in drawers or under sofas or chairs or pack them away in padded boxes if there is time. Raise objects above the floor or carry to an upper story if high water is expected. Place books in plastic bags and documents in waterproof containers. Drape furniture and other objects with plastic tarpaulins.

PLANNING

Faced with the inevitability of certain natural disasters – and the strong possibility of accidents – you would be wise to create an emergency plan. Many people outline a procedure for family safety in a crisis, but fail to use the same planning to guard their possessions. Because telephone service is often interrupted during a disaster, you should gather any essential information beforehand. Your plan will serve to minimize initial loss and to make recovery more feasible.

The plan should be simple and easy to follow. What events should be planned for? What types of belongings will require special protection? Appraise your possessions and set priorities. Consider how easily valuables can be replaced. Artworks, rare books, documents of historical or sentimental significance, and certain antique or otherwise valuable decorative arts will be all that should be considered so carefully. The plan should specify procedures to be followed in each room when advance warning leaves time for some preparation, as well as procedures for recovery after the disaster has occurred.

Planning should also include the periodic reassessment of insurance coverage. Accurate records are extremely important for preparing insurance claims. Correct inventories with descriptions and recent appraisals are very helpful. The records, of course, must also be protected from destruction in the event of disaster. Placing copies in a bank vault or other location separate from the objects is ideal.

The plan should include:

Location of valuables in your home
Location of electrical, gas, and water cutoffs

Lists noting name, telephone number, and function of federal, state, municipal, military, civil defense, and other assistance organizations

Lists of telephone numbers for local utility companies, hospitals, ambulance service, plumbers, electricians, glass companies, and other contractors and businesses

Insurance company telephone number and name of representative

Description of arrangements made in the event that possessions must be relocated or evacuated

Inventory of resources needed in an emergency. Give location of items to be stockpiled and what each item is to be used for

List of conservators, specialists, and professionals who may be called on to give expert advice and guidance

Local museums can help you with names of conservators. To plan for a widespread disaster when local conservators might be too busy for assistance, get the names of out-of-town conservators to consult via long distance. **Watch out for amateurs!** Check training and credentials until you are confident that the person you have decided to consult will recommend treatment of the highest quality according to established standards. Review the lists periodically and update any information that might have changed, such as individual names, addresses, and telephone numbers.

SAMPLE EMERGENCY PHONE LIST

Date of completion or update of this form

Service	Company and/or contact	Phone Number
Fire Department		
Police		
Sheriff		
Ambulance		
Hospital		
Civil Defense		
Utility companies		
Security service		
Insurance company		
Attorney		
Electrician		
Plumber		
Carpenter		
Exterminator		
Cold storage and/or freezing facility		
Locksmith		
Architect or builder		
Janitorial service		
Glass repair		
Photographer (for insurance purposes)		

SAMPLE LIST OF SUPPLIES

Keeping some supplies on hand for emergency use is a good idea. Space limitations will, of course, determine the amount and variety you can store. Your list of supplies, including a location inventory of supplies on hand, should be included with your disaster plan. Protect these supplies and check them periodically so that they will be usable during a storm or other disaster. Metal garbage cans are useful for stockpiling some items so that they will be easily retrievable in an emergency. Sources should be listed for all other essential supplies.

batteries
boxes (cardboard cartons and plastic milk crates)
cleaning supplies (brooms, sponges, mops, detergents, etc.)
clothesline (30 lb.) and plastic clothespins
dehumidifiers
disinfectant
extension cords (3-wire, grounded)
fans
freezer paper
hygrometers (humidity indicators)
labels
ladders
lights (chemical light sticks, incandescent work lights, flashlights)
paper towels
plastic milk crates
plastic sheeting
plastic trash bags
plywood
portable generator
protective clothing (masks, gloves, boots, coveralls)
scissors
tape (packing and duct)
tools (crowbar, hammer and nails, hand saw, screw driver)
towels and cloth diapers
unprinted newsprint
water hoses
water vacuum
waxed paper

RECOVERY

Should the worst happen and you do experience a disaster, you must immediately decide what can be salvaged. Recovery efforts will be determined by the type and extent of the damage.

The first and most important step is to consult a conservator whose expertise includes the type of item which has been damaged. In dealing with mechanical damage such as breakage, punctures, or tears, you can gently place the object and any pieces in a suitable container and take it to a professional. Make sure that you keep vibration at a minimum during transportation. In the case of fire, smoke deposits carbon residue in a greasy film over all surfaces of an object. Only a conservator can safely clean this film without surface abrasion and permanent staining.

If an object has actually been burned, there is probably not much that can be done. In the case of water damage, however, there are various courses of action for you to take. **Remember that wet materials are extremely fragile.** Never move an object until you have gathered the materials and information needed to handle it properly. Never move an object until you have a place to put it.

The following is a brief discussion of different types of objects and procedures to follow to reduce damage. **The described procedures should be used only in the case of extreme emergency to mitigate damage until a professional conservator is available to assist or take over.**

PAINTINGS

If a canvas painting has been the victim of water damage, the wet canvas will expand, causing stress to the paint layers. If the canvas dries out too quickly, the paint will crack and flake. Try to obtain professional help as quickly as possible, because mold and mildew will start to grow. If there is no chance of immediate help from a professional, a wet painting that has suffered no structural damage can be unframed and placed face down on a waterproof surface, such as a table covered with plastic sheeting, and gently blotted from behind. It should be moved from the wet area and allowed to air dry in a stable environment, that is, an environment free from fluctuations and extremes of temperature and humidity.

Flood water carries mud, oil, salts, and other chemicals that will be absorbed by the canvas. If clean water is available, **gentle** rinsing may be done. Do not submerge. **It is essential to be sure that the paint layer is secure.** Soft-bristle artists' brushes may be used to help dislodge mud, but never rub the painting with hands or any other apparatus.

If the painting's surface displays any signs of physical damage, such as cracking, flaking, or dissolving, it cannot be laid face down. Place it face up on a table to dry. Carry it face up also, so as not to disturb the loose paint. Take the painting to a conservator as soon as possible. It may be wrapped in glassine or polyethylene for transport. Glassine is a translucent, smooth paper used by museums to wrap certain objects because of its resistance to adhering even under moist conditions; it is available from conservation supply sources and most art supply stores. The conservator may be able to explain how you can apply a "face" of tissue to the surface to preserve the paint in place until you can get it to the studio.

Use these same procedures for a painting on wood ("panel" painting). Damp or wet wood will expand and stress the paint layers on top. It must be dried very slowly under the watchful eye of a conservator to save the image and keep the panel from splitting.

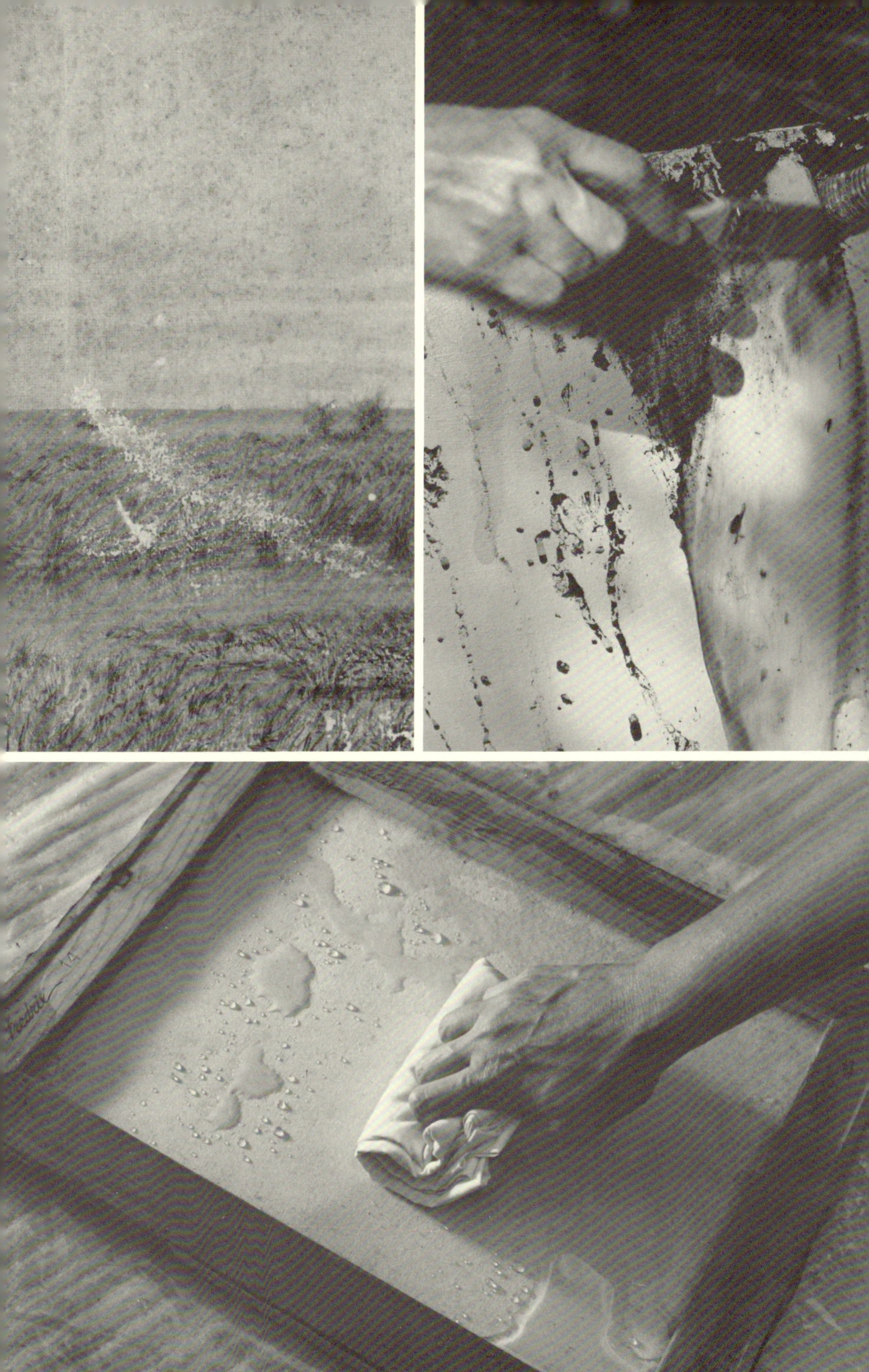

Above left, canvas with flaking paint, *right*, brushing mud from a canvas painting; *below*, blotting water from the back of a painting

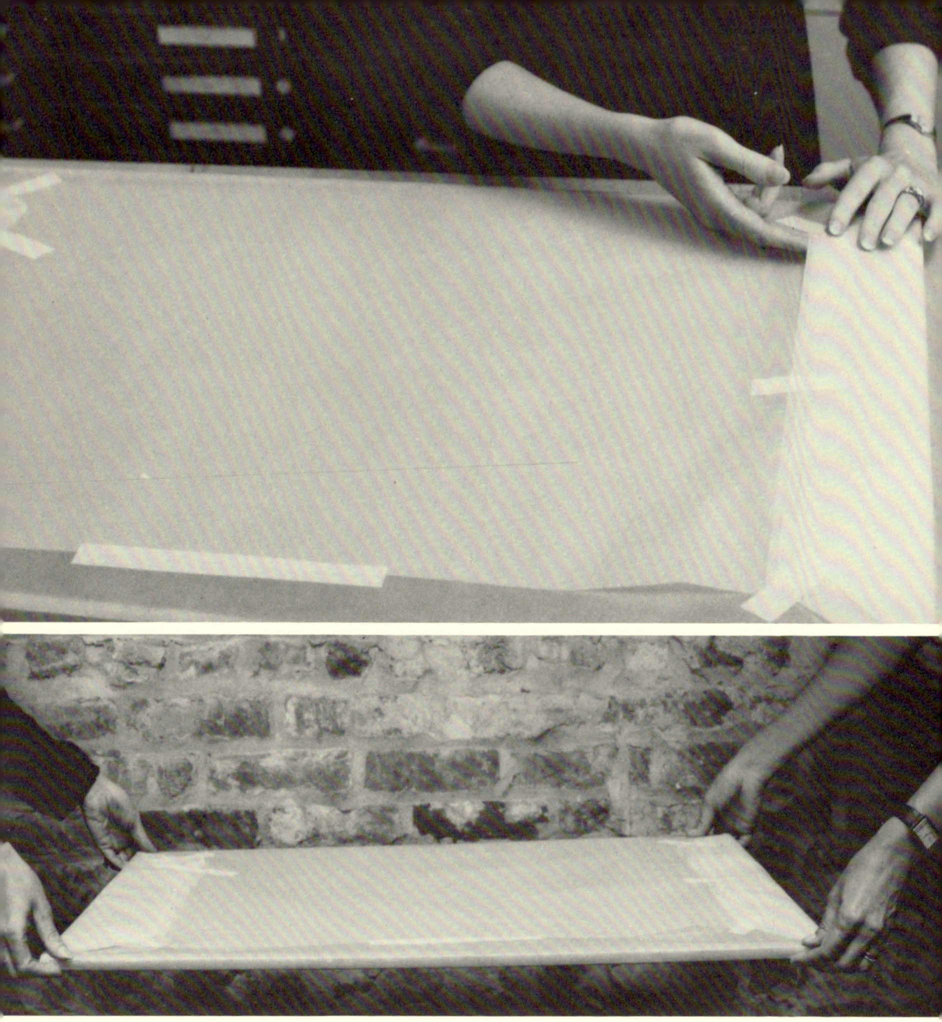

Above, wrapping a painting in glassine; *below*, carrying a wrapped painting

If a painting suffers physical damage, such as tears or gouges but is not water damaged, you can wrap it with glassine and transport it fairly safely. **The important thing to remember is to avoid further vibration**. Do not touch any part of the canvas if you can help it. A conservator can consolidate the loose paint, mend any tears or punctures, or provide a new lining for the painting without disturbing the artist's work.

PAPER AND BOOKS

Prints, drawings, documents, and books all contain paper as their basic support. If such objects have been casualties of a flood and have survived the soaking intact, recovery may be possible. Wet books will warp and swell, however, and no amount of careful drying will completely eliminate these structural changes. Essentially, paper will react to high humidity or water in the following ways: inks and paints may blur and run, adhesives will weaken or give, dimensions may change, and the paper will also become stained. Pieces of stacked, wet paper will probably stick together, particularly if the paper is coated. Magazines and art books, for example, are generally printed on coated paper. Mold growth and insect infestation are also dangers.

Because mold starts to grow on wet paper objects within 48 to 72 hours, temperature and humidity should be controlled as soon as possible. Try to get an air conditioner running in summer. Set up fans to keep air circulating. Remove any wet material that may contribute to keeping the humidity high, such as waterlogged carpet or trash. If possible, move the damaged items to a work area with better environmental conditions.

Freezing is recommended for the recovery of large numbers of wet books and documents, particularly if they are thoroughly soaked. With planning, a local cold-storage locker or a professional disaster recovery service may be identified as a potential source of help for large or bulky collections, while a conservator can recommend a safe temperature for wet papers.

Prepare documents for freezing by placing waxed paper between the sheets. Bundles can be wrapped in freezer paper. The papers must be dried under the direction of a specialist. Freezer drying, vacuum thermal drying, or vacuum-freeze drying are the preferred techniques for frozen documents or books. Books can be prepared for freezing by wrapping loosely in freezer paper and

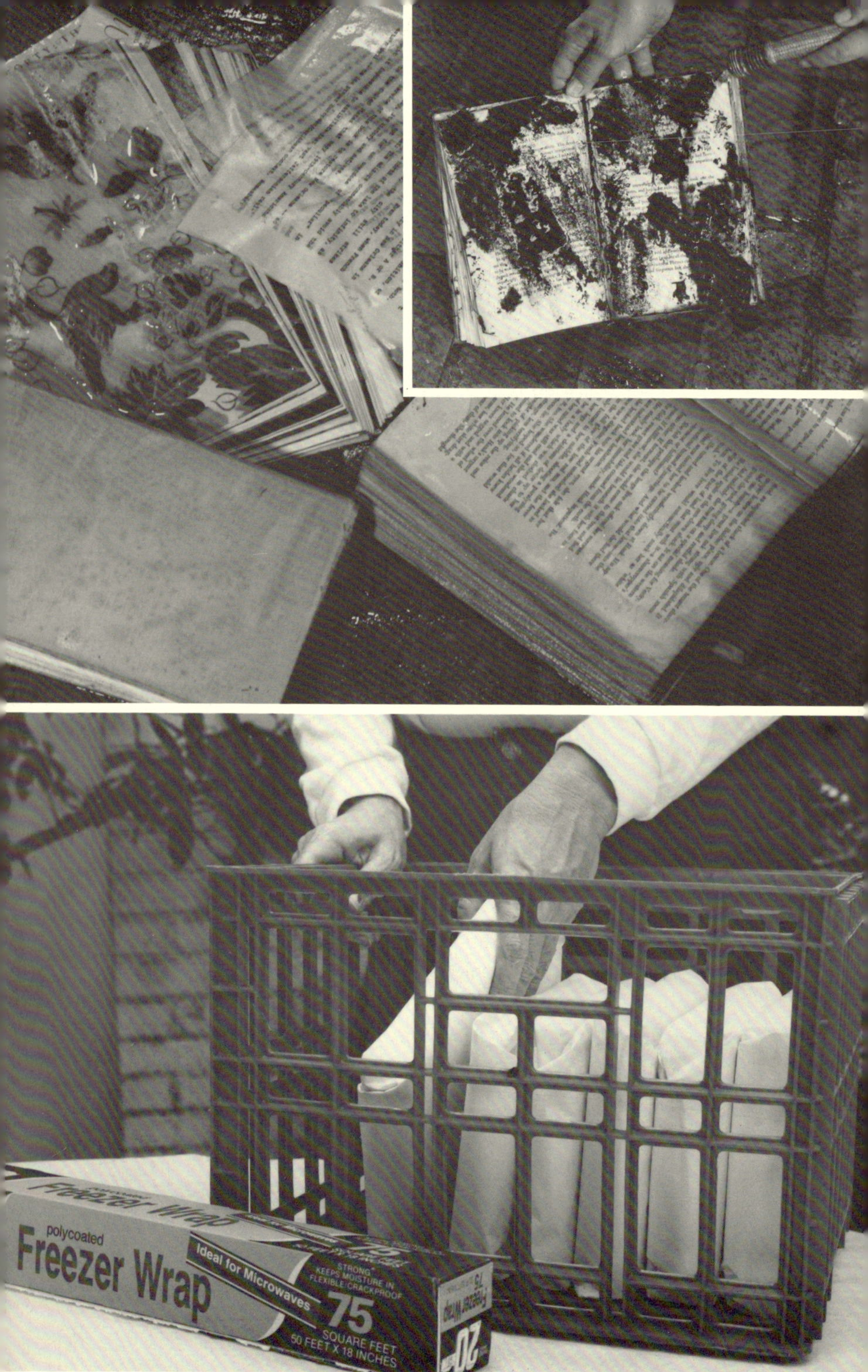

Above, water-damaged books and documents, *inset*, rinsing mud from a book; *below*, packing damaged books for freezing

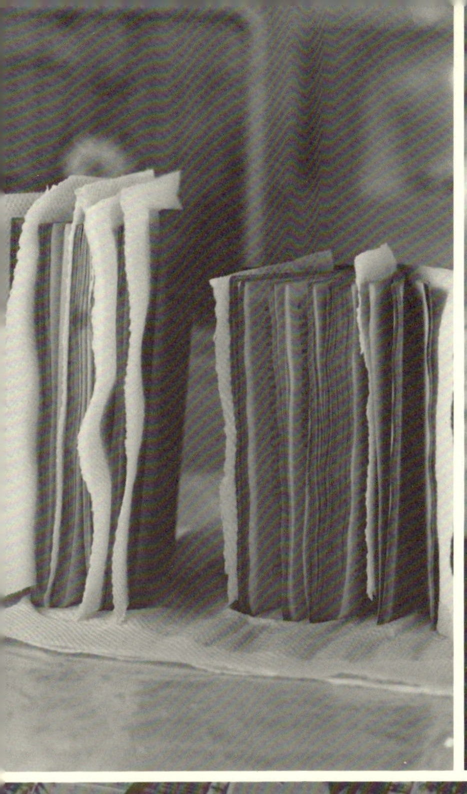

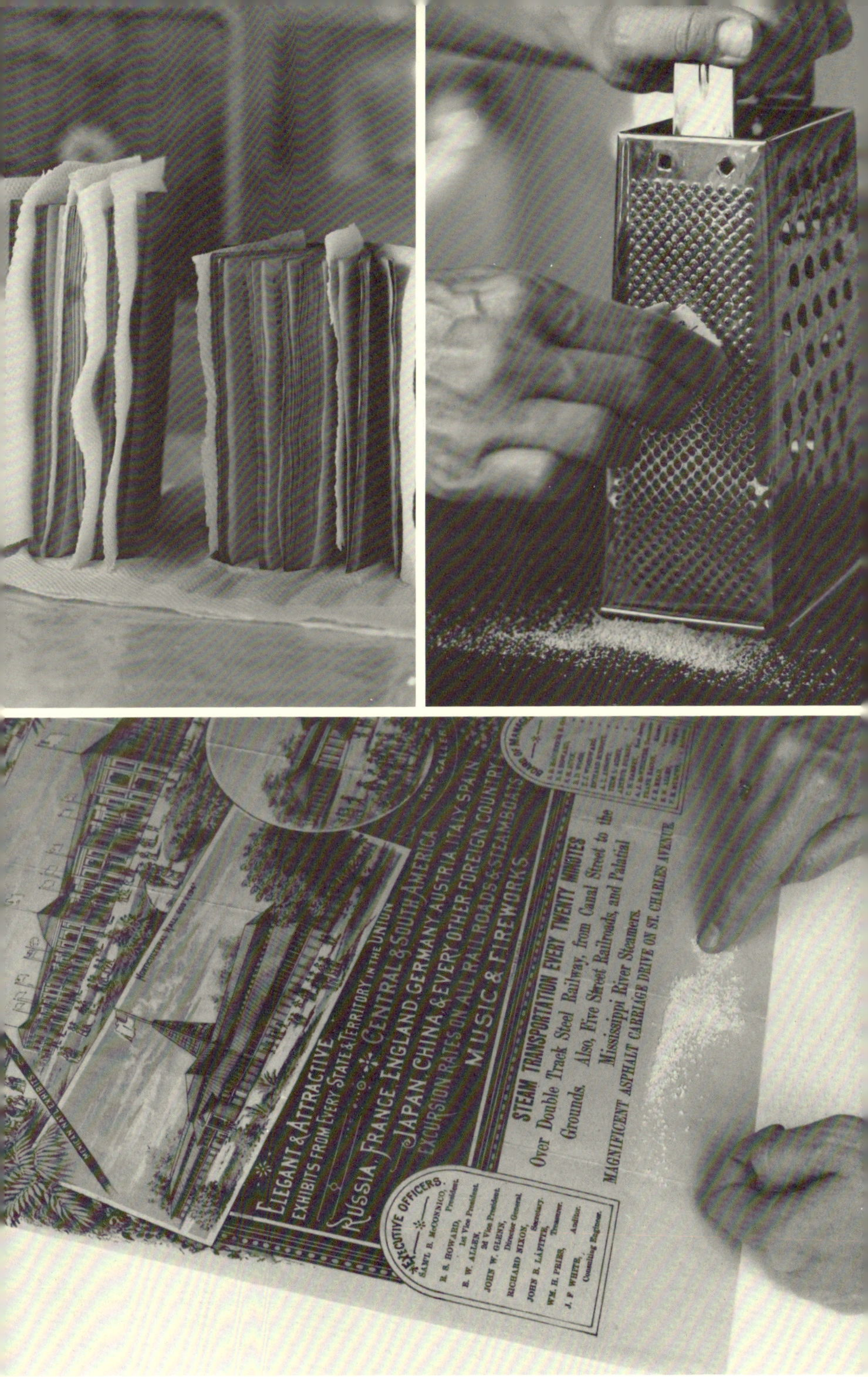

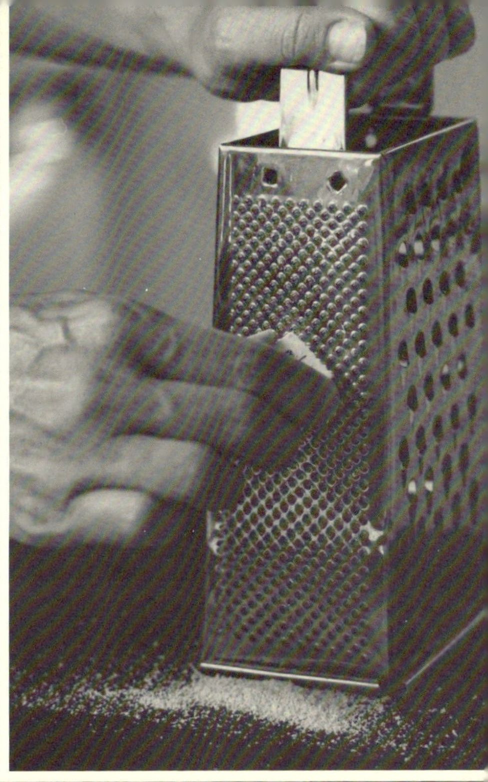

Above left, air drying water-damaged books, *right*, making powdered eraser; *below*, cleaning paper with powdered eraser

packing spine down in boxes. Plastic milk crates are ideal for transporting books since the crates provide a strong, ventilated support not susceptible to deterioration from water.

If freezing is not feasible, air drying must be started as soon as possible. Never try to use an iron to hasten drying or try to unframe a paper object that has adhered to glass. Secure a clean, dry environment where the temperature and humidity are as low as can be managed. Prepare the drying area in the same manner as described above with open windows and fans to circulate air continuously. Place each single sheet face up on a clean, flat surface if space permits. If necessary, small stacks may be made with white paper towels or unprinted newsprint as interleaving.

If you are drying books, you may stand them on end if the covers are intact. Place them on a clean, flat surface to allow the water to drain. Paper towels may be placed between the covers and the flyleaves at this stage. Change these towels and those under the books frequently. After books have begun to dry, gently begin to interleave the pages with white paper towels or unprinted newsprint. Fan the book open slightly and place a sheet between every 20 to 50 pages. Replace the wet inserts as often as possible, placing dry ones in different locations within the book. If pages are stuck together, do not force them apart. When the pages feel dry in the middle of the volume, remove the interleaving and allow the book to continue air drying for several more days.

Any document or book printed on coated paper or any object with running or blurred ink should be frozen immediately.

Avoid overexposure to ultraviolet light. Paper will become brittle and deteriorate and inks and paints will fade. Although some sunlight may be recommended for killing mold or mildew, you should guard against extended exposure to sunlight or fluorescent light in the recovery period. If you find mold growth on dry paper, it may be possible to brush off the powdery accretion with a soft brush. Do not attempt to treat artworks done with media that might smear. Under moist (not wet) or dry conditions, some exposure to sunlight may be helpful to kill the mold, but only if the object has not already suffered light damage. A professional should advise you about attempting treatment yourself; instead, he or she might treat the object with chemical fumes to kill the mold.

Mud and sewage can be **gently** rinsed off a paper object while it is still wet. If dry dirt has been deposited on the paper object, it may be possible to clean it using a soft camel's-hair brush or a powdered eraser (see instructions below). Be careful to avoid the design areas since most media will smear or fade. Be particularly careful with pastel or charcoal drawings as well as oil pastel works and crayon drawings. Only try to clean paper that has a firm surface.

Powdered eraser can be made by using a white vinyl eraser and shredding it into crumbs with a cheese grater. Pour the powder on the paper, then gently rub the powder over the surface with clean fingers or a soft brush, working on one small area at a time. The powder will loosen the dirt, and the residue can then be carefully brushed away. Conservators can reduce oil and grease stains with solvents and remove other kinds of stains and accretions by washing the paper after careful testing. They can also repair tears and holes and flatten creases and curls fairly easily. **Do not attempt any sort of major treatment yourself.**

Books, paper documents, or artworks that have suffered mechanical damage, such as punctures, tears, or broken spines, should be transported carefully to avoid further damage. A conservator can repair tears and provide a lining support for damaged pages. Bindings can also be repaired or replaced.

PHOTOGRAPHIC MATERIALS

If a disaster has affected a photographic collection, there are several steps to recovery. Dry objects are fairly stable. If negatives or slides have suffered physical damage, they can be copied by a professional. If prints have been torn or scratched, a conservator may be able to repair them. Sometimes conservators can even enhance the image in a photograph that is badly faded from excessive exposure to light.

Wet photographs, on the other hand, require a great deal of care in the recovery period. They should be dried quickly because of the danger of mold growth. In addition, water will soften photographic emulsion and cause it to swell. Once the emulsion layer has softened, dirt or other external materials, such as enclosure envelopes, may become lodged in it or cause it to be distorted. The emulsion will separate from the negative or print if allowed to remain wet for more than a few days. Some of the dyes in color photographs can disappear altogether and the color layers may separate. Water may affect the varnish coating on cased photographs, such as tintypes and ambrotypes, or it may corrode the copper base of silver-coated daguerreotypes.

Air drying is acceptable for photographs if they can be laid out in single layers without stacking. They can be placed on blotters or nylon screen. Wet black-and-white prints or negatives that have been stacked or stored in an enclosure must be separated before drying. If gentle pulling disturbs the emulsion, do not continue trying to separate. Submerge the materials in cool, clean water, watching carefully for evidence that the emulsion may be dissolving. If the photographs float apart, they may be air dried. If it is not possible to separate photos easily, they should be removed from the water and frozen. A conservator should advise you on procedures for freezing and for transporting the wet materials.

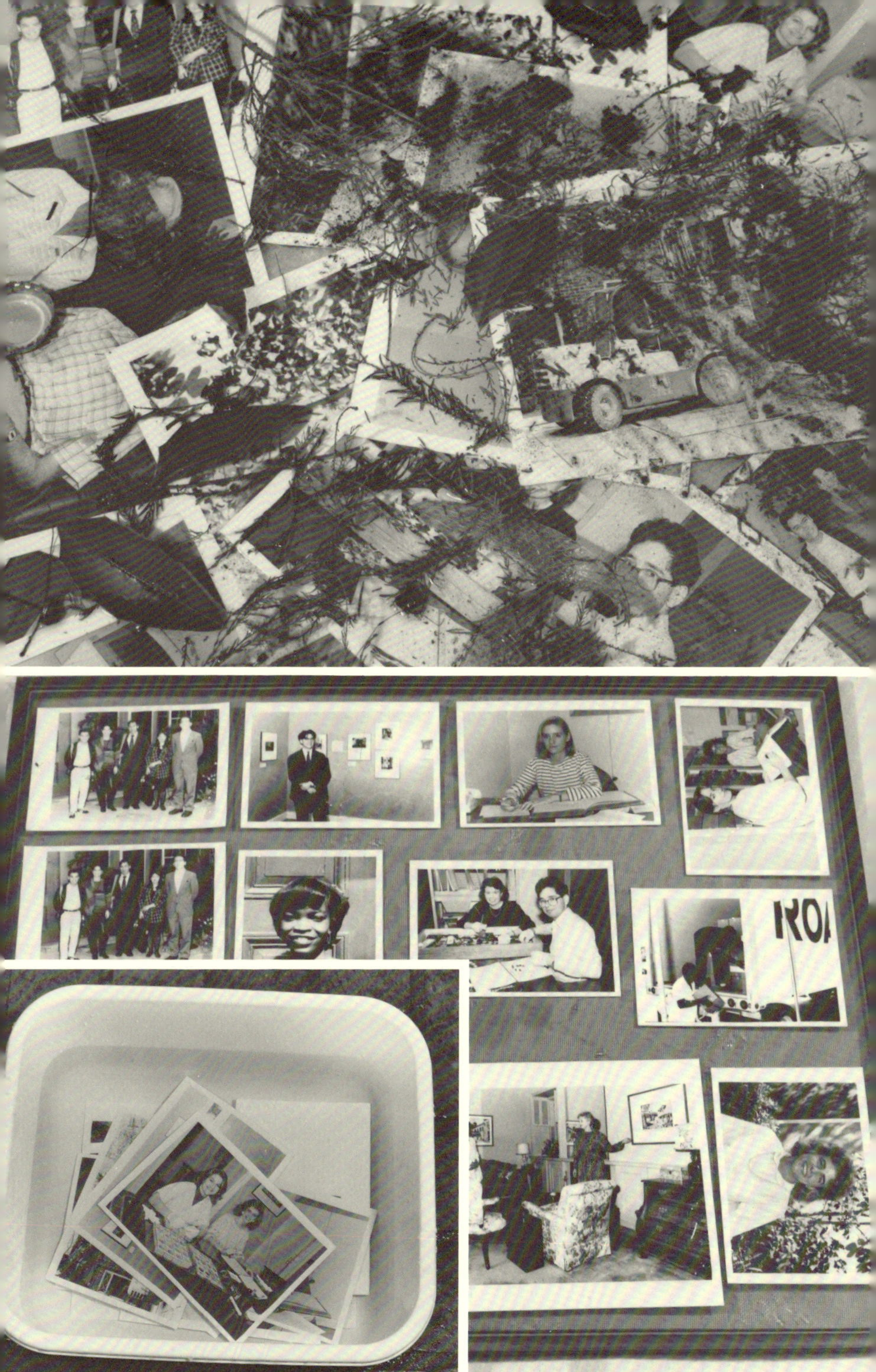

Above, water- and debris-damaged photographs; *below*, drying photographs on a screen, *inset*, soaking photographs to separate

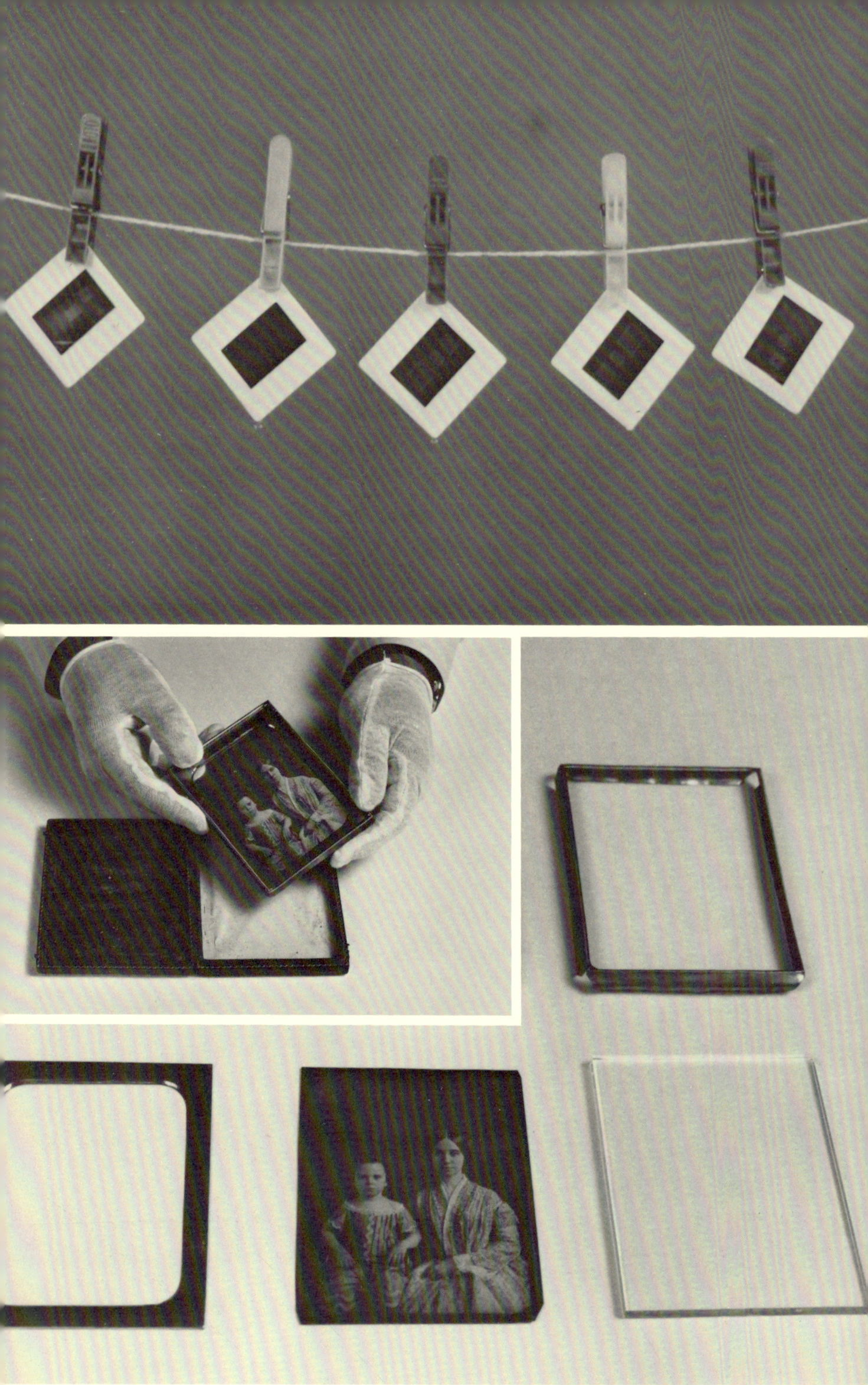

Above, slides drying on a line; *below*, drying a daguerreotype, *inset*, removing a daguerreotype from its case

Wet color materials may be handled the same way, but they are much less stable than black and white. Kodak Photo-flo or another similar commercial photographic wetting agent may be used to rinse slides, which should be supported upright to dry, either propped at an angle or hung on a line.

Cased photographs can be air dried. Daguerreotypes should be removed from the case and placed face up on blotters along with the parts of the case. Likewise, tintype and ambrotypes or glass-plate negatives should be dried emulsion side up. **Do not touch the surface of these photographs.**

TEXTILES

In a disaster, textile items are most likely to suffer from water damage, mud, dirt, or tears. Like paintings and paper, textiles are subject to rapid mold and mildew growth under wet conditions and in an environment of high humidity. Staining and deterioration of fibers is the result. If water damage has occurred, you may clean and dry the objects as soon as possible but without using extreme measures. Strong sunlight, high heat, and irons should be avoided. Laying an object out on a polyester screen indoors where the humidity is low is ideal for drying fragile or old fabric. Never hang wet woolens or old silk.

If a fragile fabric has survived water without shrinking or without its colors running, it can be gently washed, avoiding any agitation, in cold distilled or deionized water with a non-ionic detergent, such as Orvus. Washing should be done before the object dries, if possible. Rinse thoroughly. The object may be supported by a polyester screen during the washing and rinsing process. If a dry object is to be washed, dyes should first be tested for colorfastness by saturating a small area with water and blotting it with clean, absorbent paper. If color from the dye appears on the blotter, do not wash the textile in water. The same test

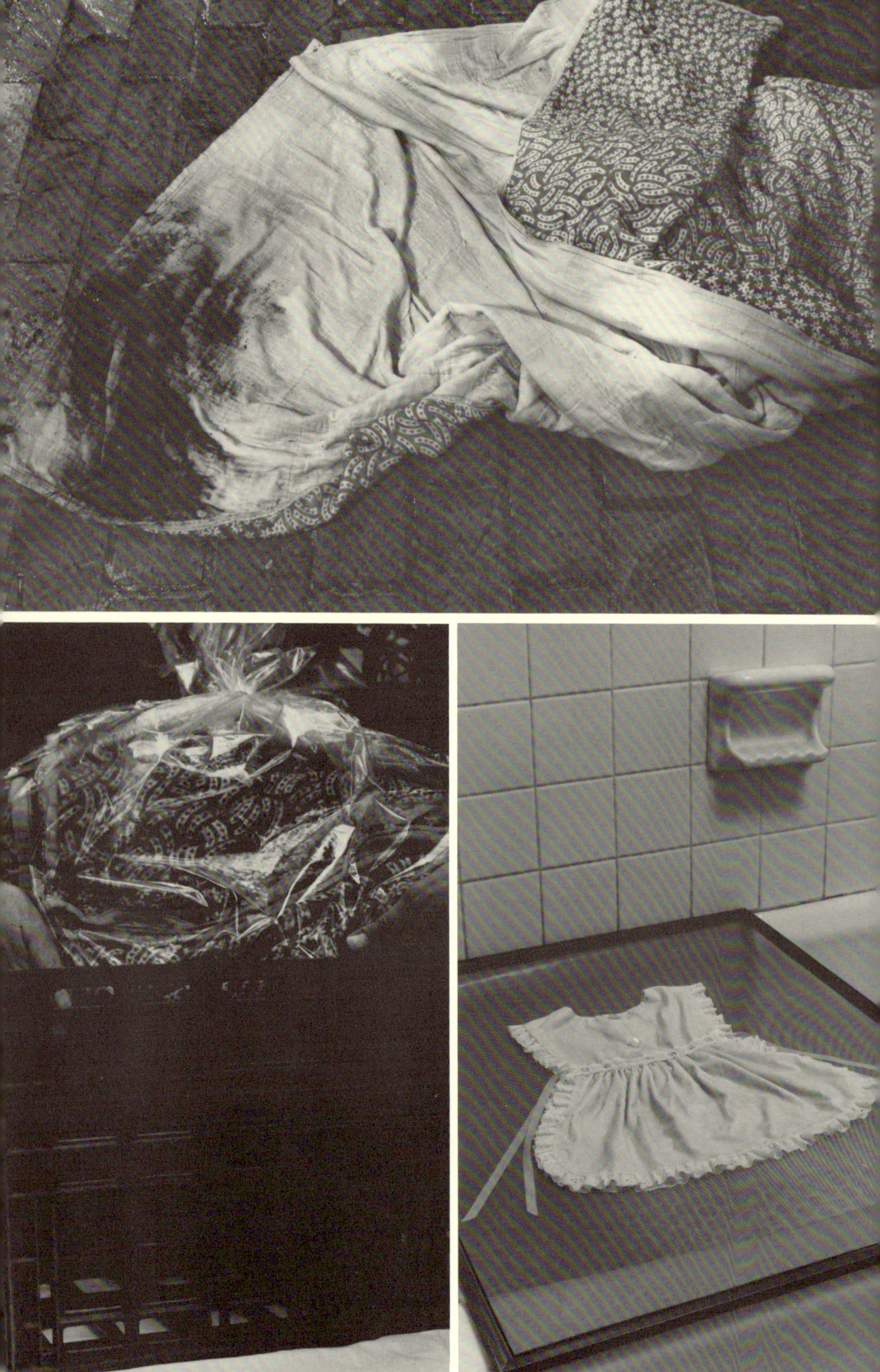

Above, water- and mud-stained quilt; *below left*, transporting damaged textiles for freezing, *right*, drying wet fabric on a screen over bathtub

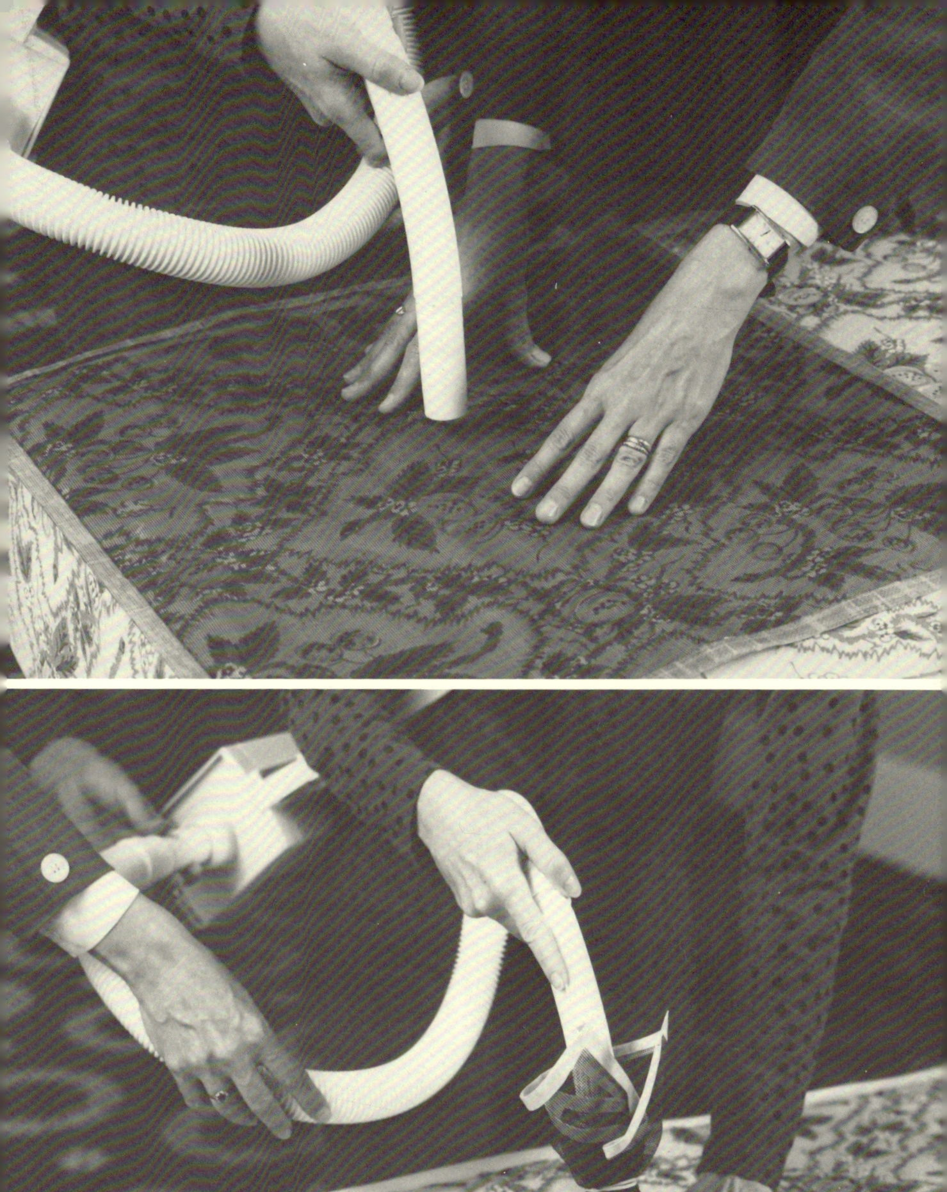

Two methods of vacuuming fabrics

should be done for the detergent. Refrain from using commercial laundry detergents or bleaches for old or fragile fabrics because they will deposit damaging substances in the fibers.

Wet textiles can be frozen to prevent mold growth if cleaning and drying cannot be performed quickly; drying can then be done later when an expert can control the process. Freezing also prevents some stains from setting and stabilizes bleeding dyes. For freezing, textiles can be placed loosely and gently in polyethylene bags. Milk crates or plastic bread trays are ideal for carrying bundles. Very low temperatures will ensure that the ice crystals are small.

When dry fabrics have deposits of loose dirt, it may be possible to perform gentle vacuuming. A fine plastic screen held over the nozzle of a vacuum cleaner on its lowest speed will remove the dirt and avoid stress to the fabric. If further cleaning is required, request hand dry cleaning from your cleaner.

It is important to be aware of basic considerations for handling delicate fabrics. If a valuable object has sustained tears, or if the fibers are extremely weak or brittle, a textile conservator should be consulted.

FURNITURE AND WOODEN OBJECTS

If wood has become wet or waterlogged, it must be dried very, very slowly. Remove the object from standing water and consult a professional immediately. If wood is unpainted, mud can be removed by gently and carefully rinsing. A soft brush can be used to dislodge the mud, but it is advisable not to use any other kind of tool. For painted wood, rinse only if the paint layer can be determined to be completely stable. Upholstery and metal attachments should be removed and treated according to guidelines found in the sections on textiles and metals.

Drying is a delicate matter and should be done with the supervision of a conservator. Objects can be draped with polyethylene to keep them from drying out until an expert can be called in. Drying should be done under very controlled environmental conditions in order to prevent warping and severe cracking. The process can take several months. Surface water can be blotted dry, but be absolutely certain that the water has not been absorbed into the piece before assuming that it is fully dry.

Cleaning a deposit of dust and debris can be done carefully with the brush attachment of a vacuum cleaner. A small piece of plastic screen can be held over the nozzle to avoid sucking up any furniture marquetry or veneer that may not be obviously loose. If loose ornamentation is evident, do not try this procedure.

If the dirt on dry furniture is heavy, mineral spirits may be used as a cleaner, but you should first perform a spot test because some finishes are soluble. To test, moisten a cotton swab and dab an inconspicuous spot. If the spot becomes whitish or sticky, do not continue. Consult a professional if there is the least chance of removing an irreplaceable finish. A painted wooden object cannot be cleaned safely by anyone except a conservator.

Determination of suitable chemical treatment for insects and fungi may be made with the advice of a professional pest-control service, but you must also consult a conservator. The conservator will help you determine which chemicals are safe for the various components of your object. For instance, some chemicals may corrode the metals in furniture hardware.

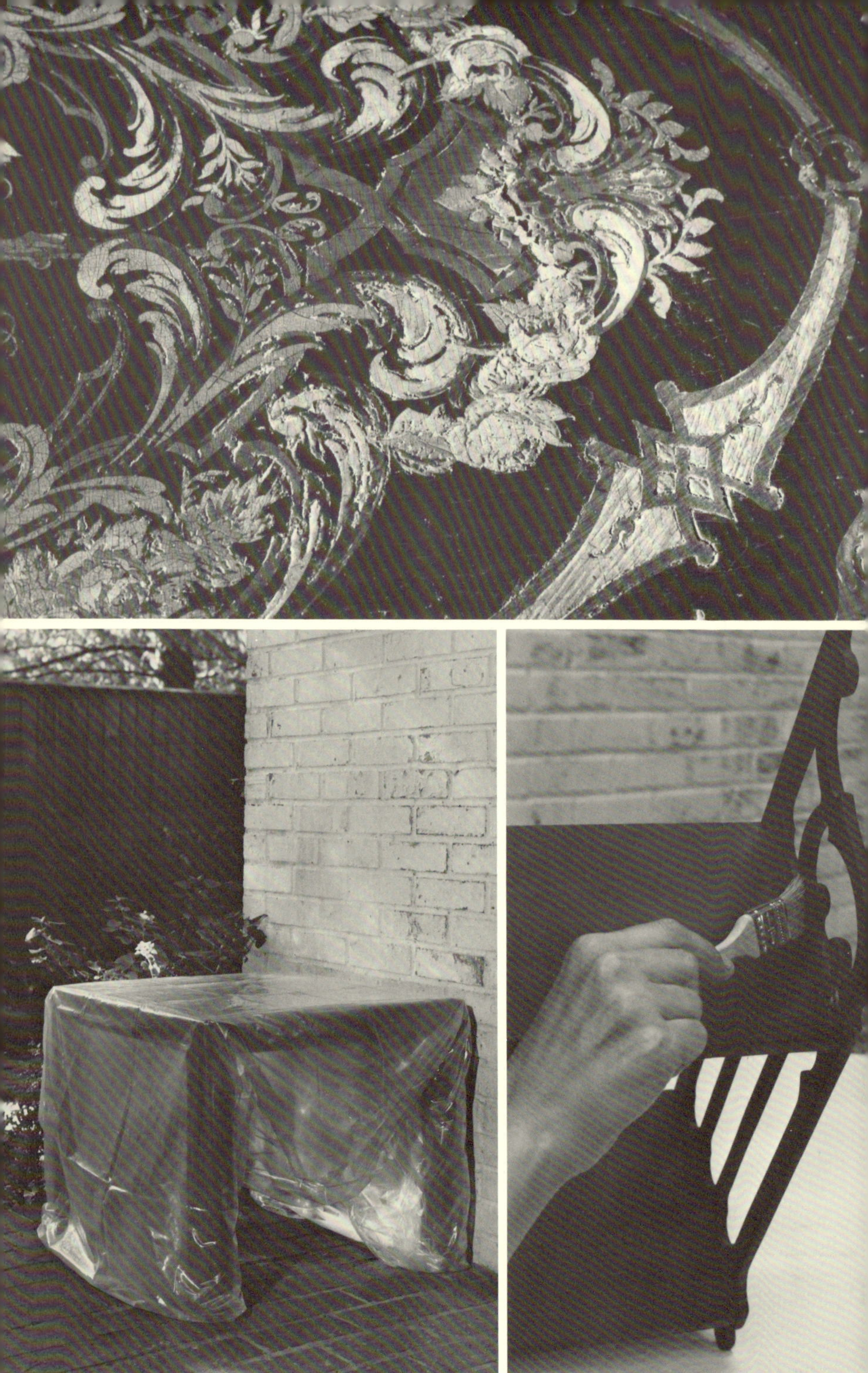

Above, damaged furniture inlay; *below left*, table draped in polyethylene to prevent premature drying, *right*, brushing dirt from furniture

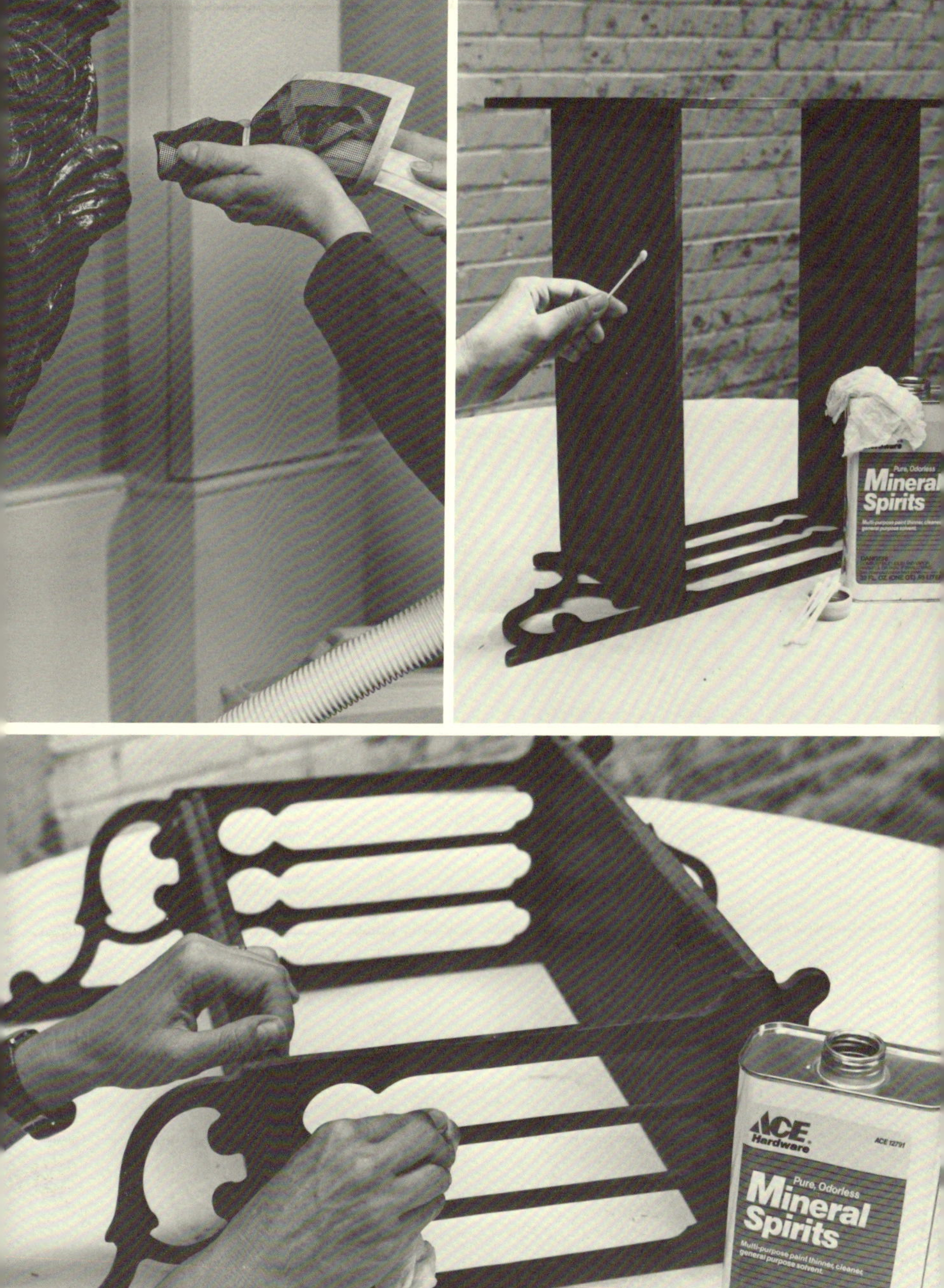

Above left, vacuuming dust and debris from furniture, *right*, testing solubility of furniture finish; *below*, cleaning dirt from furniture with mineral spirits

METAL OBJECTS

The most preventable or reversible damage that may affect metal objects is corrosion. Corrosion appears as rust or tarnish. Iron is the metal most vulnerable to rust, but others, including copper, brass, and bronze, are susceptible as well. Rusting occurs quickly and may be hastened by salts carried in water or found in the air under certain atmospheric conditions. Damaging salts are also found in human perspiration. Pure gold will not be affected by moisture alone, but if it is mixed with an alloy, the object may be affected by rust or tarnish. Tarnish on silver comes from exposure to chlorides and sulphur in the air and water. Chlorine in city water will cause significant problems if susceptible metals are left wet for very long.

Mud should be rinsed from wet metals with fresh, clean water. Dry them as quickly as possible. A very soft cloth may be used on anything that does not have loose pieces. Especially for cast iron, a blow drier may be used, or the object may be placed in an oven overnight. Oven temperature should be set low, at 150°F, if possible; 200° may be too high for soldered pewter or silver. Of all metals, iron should be returned to an atmosphere of low relative humidity. A constantly maintained level of 40% or 50% is ideal. A professional should examine any valuable piece that has been exposed to potentially damaging moisture.

Once the metal object is dry, dirt or oils may be removed with alcohol or mineral spirits. Very light rust may be rubbed from iron with fine bronze wool. You should be aware that once the surface has been cleaned, it is vulnerable to new rust. Consult a conservator for protective coatings and for treatment for more serious corrosion problems. Protect the object from high humidity, industrial pollution, corrosive salts, acid conditions, and fingerprints. Strong acids and commercial rust removers should never be used on valuable objects since they may cause extensive pitting of the metal.

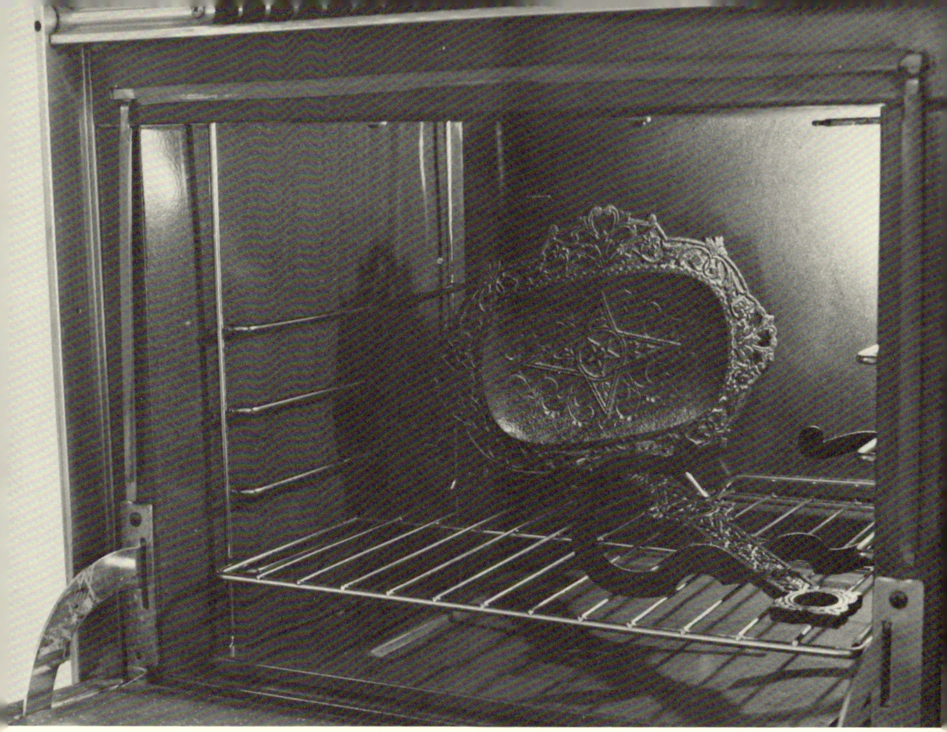

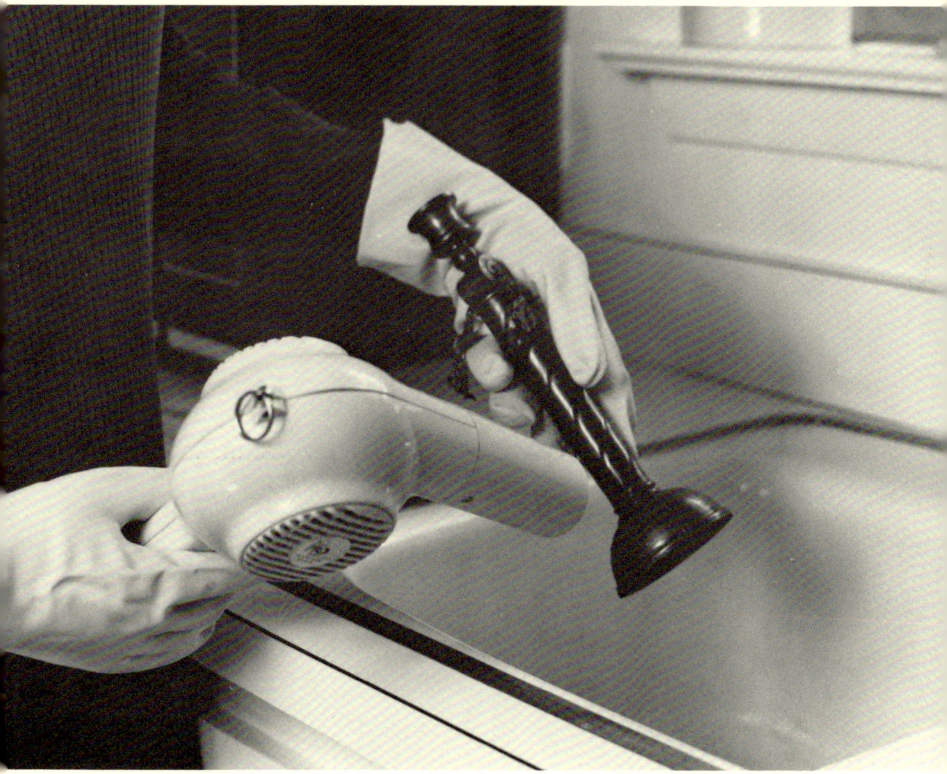

Above, drying cast iron in an oven; *below*, drying metal candlestick with a blow dryer

GLASS AND CERAMICS

If a glass or ceramic object has been affected by debris or dirt deposits, cleaning methods should be considered carefully. The method will depend on whether the object is porous or nonporous. Nonporous objects such as high-fired ceramics and glass may be washed with a mild soap such as Ivory. Avoid ammonia, strong detergents, and scouring powder. Since detergents leave a residue, it may be necessary to rinse the object with distilled water. Use only lukewarm water for washing and rinsing. Excessively hot water may cause certain glazes to crack or craze. Great care should be taken to prevent a plate or a glass from slipping from your hands. Use a plastic dishpan inside your sink. **Never wash valuable objects in a dishwasher**. Chemicals in dishwashing detergent and the high heat may remove overglazed decoration such as gold rims or patterns.

Glazed porous wares such as majolica, faïence, and creamware can be cleaned with a wet cloth. Do not immerse. Unglazed porous wares such as Indian pottery must be dry-cleaned by a conservator. Dust may be removed with a soft artist's brush. Porous objects affected by water should be taken to a conservator. Undesirable elements carried in flood water may be absorbed into porous wares causing staining and degradation of the glaze.

If a glass or ceramic object is broken, carefully gather the pieces and place them in a padded container. Small pieces can be collected in an egg carton so they will not rub against each other. Do not try to fit the pieces back together because the broken edges will abrade each other, resulting in loss of material. Let the professional repair broken objects: a conservator will know the adhesive that is best for the particular kind of ceramic or glass.

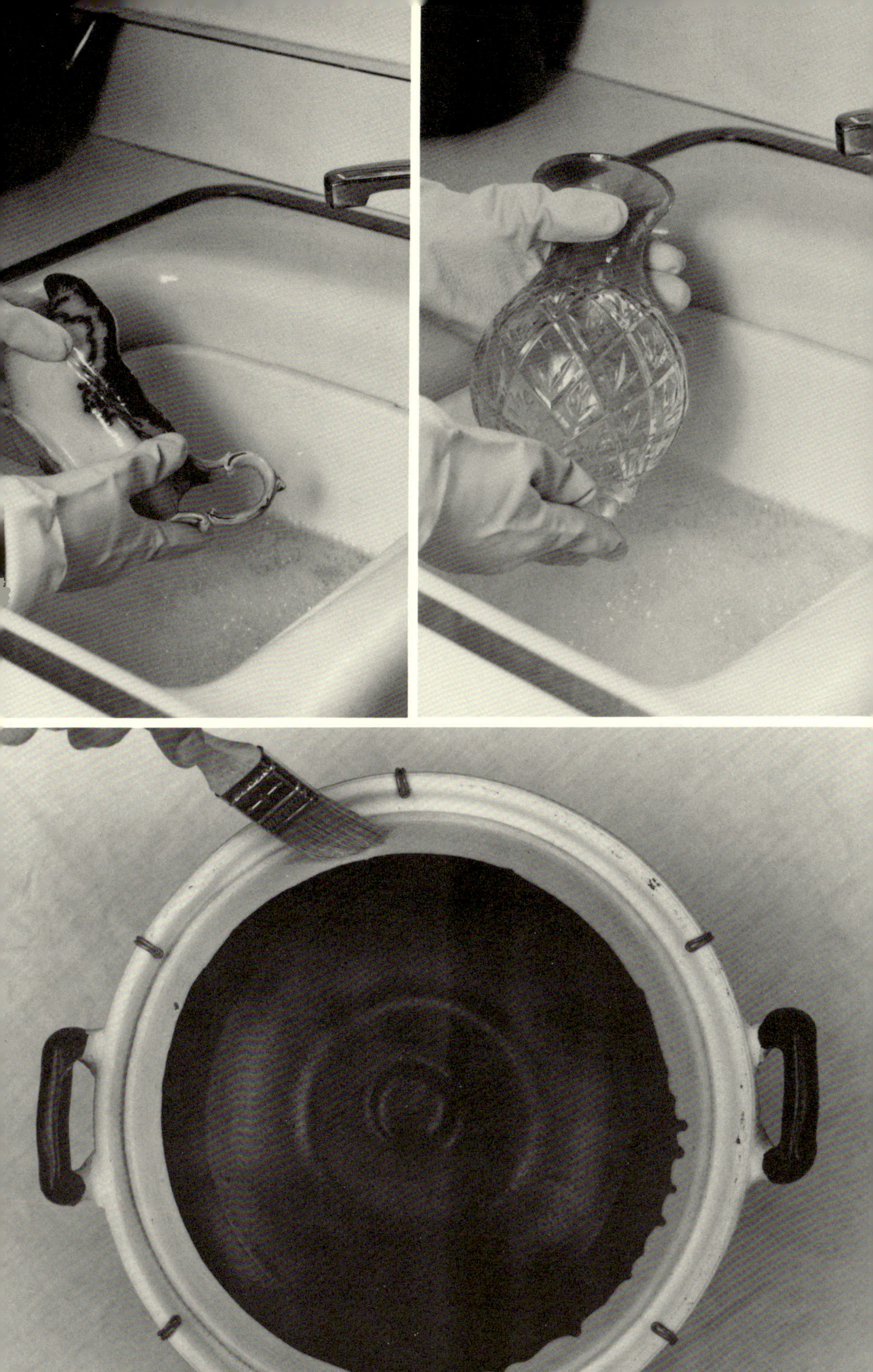

Above, washing ceramics and glass in a plastic dishpan; *below*, removing dust from unglazed porous ware with a brush

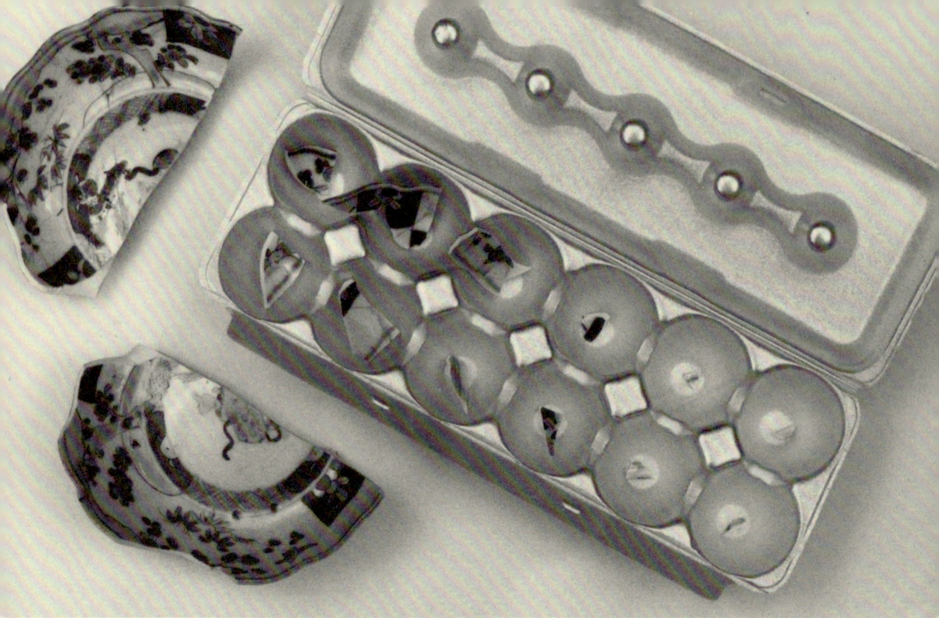

Pieces of broken ceramic placed in an egg carton for transport

The costly destruction of Hurricane Andrew in August 1992 put many emergency plans to the test. Those in the path of this devastating storm may have benefitted from good planning. Undeniably, there are times when no amount of preparation will save lives, much less homes and belongings. It is certain, however, that a heightened awareness of potential danger and a general awareness of where and how to obtain help will lessen risk and improve chances for recovery.

Although it is impossible to foresee every disaster, a general assessment of danger is the first step to preparedness. After determining the most likely perils, prevention measures can be planned and implemented. A written plan including the names and telephone numbers of all those individuals and organizations available to help can be carefully worked out. A general awareness of methods of recovery is essential. Above all, when planning, consider the conservator your most valuable source of aid.

WHERE TO FIND CONSERVATION INFORMATION

In addition to lists of conservators maintained by local museums, the American Institute for Conservation of Historic and Artistic Works, 1400 Sixteenth Street, N.W., Suite 340, Washington, DC 20036, is available to assist in finding the right conservator. The AIC regularly publishes a directory of members by location and by specialty. Other organizations that can provide conservation-related information are:

American Association for State and Local History
172 Second Avenue North, Suite 202
Nashville, TN 37201

American Association of Museums
1225 Eye Street, N.W.
Washington, DC 20005

American Library Association
50 East Huron Street
Chicago, IL 60611

American Society of Appraisers
535 Herndon Parkway, Suite 150
Herndon, VA 22070

Association for Preservation Technology International
P. O. Box 8178
Fredericksburg, VA 22404

National Fire Protection Association
1 Battery March Park
Quincy, MA 02269

National Institute for the Conservation of
Cultural Property
3299 K Street, N.W., Suite 403
Washington, DC 20007

National Park Service
P. O. Box 37127
Washington, DC 20013-7127

National Trust for Historic Preservation
1785 Massachusetts Avenue, N.W.
Washington, DC 20036

Society of American Archivists
600 South Federal Street, Suite 504
Chicago, IL 60605

BIBLIOGRAPHY

Buchanan, Sally A. *Disaster Planning: Preparedness and Recovery for Libraries and Archives*. Paris: UNESCO, General Information Programme and UNISIST, 1988.

Burke, Robert B., and Sam Adeloye. *A Manual Of Basic Museum Security*. Leicester, Great Britain: International Council of Museums, 1986.

"Disaster: Preparedness and Response." *OJO: Connoisseurship & Conservation of Photographs* (Summer 1992): 1-2.

"Emergency Salvage Advice for Photos," *Technical Leaflet*. Andover, MA: Northeast Document Conservation Center, August 1990.

Jones, Barclay G. *Protecting Historic Architecture and Museum Collections from Natural Disasters*. Stoneham, MA: Butterworth Publishers, 1986.

Levenstein, Mary Kerney, and Cordelia Frances Biddle. *Caring For Your Cherished Possessions*. New York: Crown Publishers, 1989.

MacLeish, A. Bruce. *The Care of Antiques and Historical Collections*. 2d ed., rev. Nashville, TN: American Association for State and Local History, 1985.

Meister, Pamela, ed. *Southeastern Museums Conference 1991 Disaster Preparedness Seminar Proceedings*. Baton Rouge, LA: Southeastern Museums Conference, 1991.

National Committee to Save America's Cultural Collections, Arthur W. Schultz, Chairman. *Caring for Your Collections*. New York: Harry N. Abrams, 1992.

Piggott, Tami L. *Disaster Preparedness: How to Develop and Implement a Plan*. Unpublished paper presented at the Southeastern Museums Conference, October 1990.

Solley, Thomas T., et al. *Planning For Emergencies: A Guide For Museums*. Washington, DC: Association of Art Museum Directors, 1987.

Walsh, Betty. "Salvage Operations For Water Damaged Collections." *Western Association for Art Conservation Newsletter*. 10 (May 1988): 2-5.

Waters, Peter. *Procedures for Salvage of Water-Damaged Library Materials*. 3d ed. Washington, DC: Library of Congress, 1988.

Williams, Marc A., and Heidi Miksch. *Hurricane Handbook*. Unpublished booklet prepared by the Winterthur Conservation Program, August 1977.

SELECTED TITLES FROM THE HISTORIC NEW ORLEANS COLLECTION

Preservation Guides

1. *Family Papers* by Susan Cole
2. *Photographs* by John H. Lawrence
3. *Paintings* by Priscilla O'Reilly
4. *Furniture* by Maureen A. Donnelly
5. *Books* by Pamela D. Arceneaux and Jessica Travis
6. *Matting and Framing* by Alan Balicki
7. *Silver* by Maureen A. Donnelly

Bibliography of New Orleans Imprints, 1764–1864 by Florence M. Jumonville

Boyd Cruise by Alberta Collier and Mary Louise Christovich

Crescent City Silver by Carey T. Mackie, H. Parrott Bacot, and Charles L. Mackie

Encyclopaedia of New Orleans Artists, 1718–1918 by John A. Mahé II, Rosanne McCaffrey, and Patricia Brady Schmit

Nelly Custis Lewis's Housekeeping Book by Patricia Brady Schmit

Southern Travels: Journal of John H. B. Latrobe, 1834 by Samuel Wilson, Jr.

Vicksburg: Southern City Under Siege by Kenneth Trist Urquhart

Board of Directors
Mrs. William K. Christovich, *President*
G. Henry Pierson, Jr.
Francis C. Doyle
John E. Walker
Fred M. Smith

Dr. Jon Kukla, *Director*

Dr. Patricia Brady, *Director of Publications*
Louise C. Hoffman, *Editor*
Lynn D. Adams, *Publications Researcher*

Jan White Brantley, *Head of Photography*
Judy Tarantino, *Photographer*

Maureen Donnelly, *Senior Registrar*
Maclyn Hickey, *Assistant Registrar*
Doug MacCash, *Head Preparator*
Steve Sweet, *Assistant Registrar/Preparator*

Thanks to Alison Luxner for reading and commenting on portions of this manuscript.